Elias Wildman

Instructions in the Manipulation of Hard Rubber

or vulcanite, for dental purposes

Elias Wildman

Instructions in the Manipulation of Hard Rubber
or vulcanite, for dental purposes

ISBN/EAN: 9783337223922

Printed in Europe, USA, Canada, Australia, Japan

Cover: Foto ©berggeist007 / pixelio.de

More available books at **www.hansebooks.com**

INSTRUCTIONS

IN THE

MANIPULATION

OF

HARD RUBBER OR VULCANITE

FOR

DENTAL PURPOSES.

BY

E. WILDMAN, M.D., D.D.S.,

PROFESSOR OF MECHANICAL DENTISTRY IN THE PENNSYLVANIA COLLEGE OF
DENTAL SURGERY.

FOURTH EDITION.

PHILADELPHIA:
SAMUEL S. WHITE.
1867.

PREFACE.

As the use of hard rubber as a base for artificial dentures has grown into so much favor, both with dentists and their patients, owing to its lightness, cleanliness, and perfect adaptation to the parts upon which it reposes, every operator should be in possession of the knowledge of its manipulation for this purpose. Yet, as many of the profession, from various circumstances, have not been able to obtain this information, it was believed that full and concise instructions therein would prove acceptable to such.

In working rubber, care must be exercised at every step, to produce the best results. The best material may be rendered spongy, or brittle, and worthless, by carelessness or a slovenly manner of manipulation; while, on the contrary, with care, a compact, tough, strong, and somewhat elastic compound will be produced. In the following pages the writer has detailed his own process, which, after several years' experience, has been followed by uniform success: therefore it is confidently offered to such of the profession as desire instruction upon this subject.

E. W.

Philadelphia, Oct. 1865.

CONTENTS.

ILLUSTRATIONS.

ADDENDUM.

INSTRUCTIONS

FOR

MAKING VULCANITE WORK.

A GENERAL OUTLINE OF THE PROCESS FOR MAKING RUBBER WORK.

In making this work, the impression of the mouth is to be taken in the same manner as for metal work. It must be smooth, accurate, and free from imperfections. This part of the process will not require further description. The next step is to make the model, then a model base-plate, upon which the teeth are adjusted, and with a plastic substance a model denture is built up in the form of the desired vulcanite set. This model set, placed on the model, is then enclosed in plaster in a flask, to form a mould of the model set. When the flask is separated, the teeth remain attached to the plaster in one of the sections of the flask. The model plate is removed, and grooves are cut in the plaster to allow the excess of rubber to exude from the mould. The mould is then prepared so as to prevent the rubber from penetrating and adhering to it, and then packed with unvulcanized rubber; after which the two sections are brought together under strong pressure.

It is then ready for hardening, or vulcanizing. This may

be accomplished by submitting it for a certain time to the action of hot air, steam, or hot water. Water confined in a strong boiler, called a vulcanizer, is generally used by dentists. After having been vulcanized, the set is trimmed and polished.

These several steps in the manipulation will be described in detail.

MAKING THE MODEL.

In making the model, care should be taken to have its face smooth and free from imperfections, and the body hard and compact. To insure a smooth surface from a good plaster impression, the plaster should be saturated with water. The better plan is to cast the model immediately after taking the impression: if this is not practicable, immerse it in water, to restore what is lost by evaporation. Then use such substance as the operator may prefer, to prevent adhesion. The practice of drying the impression, varnishing, and then oiling, is a useless waste of time: a thin film of oil, or a solution of soap, may be used, but neither in excess. A solution of soap can only be used upon naked plaster; if applied to varnished plaster or a wax surface, it will act upon the plaster cast into the mould, and cause the surface of the model to be pulverulent, so that the fine lines will be readily effaced. The same effect upon the surface of the model may be produced, where the impression is dry, by its absorbing the water from the cast, so as to prevent the proper chemical union taking place. To obtain solidity and strength, the plaster should be of good quality, mixed as thick as can be manipulated, and free from air-bubbles.

When the impression has been taken in wax, where there

are teeth; fill the cavities made by the teeth with water, then add a little plaster, mixed thick; jar the impression; the plaster will settle into the recess, and the water will rise; then add sufficient plaster to make the model of the required thickness, jarring to cause the plaster to consolidate better.

In parting the model from the wax impression, the better plan is to immerse the whole in water at a temperature just sufficient to make the wax plastic enough to be removed from the teeth without endangering their fracture. The advantage of warm water over dry heat is that it keeps the plaster saturated, and the wax, or wax compound, does not enter its pores; it parts freely, and leaves the surface clean, which is very important for vulcanite work. When parted in this manner, it is not necessary to oil the wax impression before casting into it.

Trim the model, and make it much thinner than for metal work. The surface must not be varnished.

MODEL BASE-PLATE AND MODEL SET.

For a base for the model set, various substances are used. Some prefer wax in sheets: these may be made by casting or rolling wax to the proper thickness, or—a better plan—by immersing sheets of paper in melted wax, withdrawing to cool, and repeating the process until the sheet has attained the required thickness. The gutta percha of commerce has been used; but it is objectionable, because when in thin sheets it is wanting in the proper rigidity, and often becomes so adhesive as to make it unpleasant and difficult to work. Others prefer a metal plate stamped up as for a metal base: block tin, or pewter, rolled rather thicker than for silver work, may be resorted to; but I prefer as a base the *pre-*

pared gutta percha; it is very plastic when heated, and possesses sufficient rigidity when cold. That prepared by the American Hard Rubber Company, I have found the most reliable.

In making a model plate of this substance, the prepared gutta percha may be softened by dry heat, or, better, in water of a temperature above 150° F. The model should be saturated with cold water. This prevents the gutta percha from adhering to its surface. The softened gutta percha is then pressed down firmly with the fingers wet with cold water, and made to fit the face of the model accurately. The articulation is procured in the same manner as for metal work. Now upon the alveolar ridge of this plate attach a strip of wax or gutta percha, to form a backing for the teeth in the model set.

Teeth and sectional blocks are made expressly for vulcanite work with long pins, or with headed pins. When teeth with plain pins are used, the pins should be curved or hooked horizontally: if the curve is made perpendicularly, and in an arc of a circle of which the base of the tooth is the centre, the pins will be easily pulled out of the rubber by a force applied to the point of the tooth.

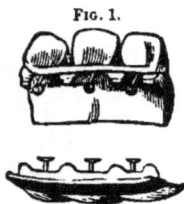

Fig. 1.

Double-headed pins are the most reliable, and, when properly made, there is no possibility of the tooth becoming detached from the rubber without a fracture of the pin or tooth.

In some peculiar cases where rubber teeth cannot be procured to answer the purpose, ordinary plate teeth may be used. They can be prepared so as to make them as firm and secure as rubber teeth, by flowing a film of gold solder upon a narrow strip of gold or platina plate. Then square

the ends of the platina pins, clamp the strip with the solder-
face, coated with borax, in contact with the pins, and heat
up to the fusing-point of the solder: this

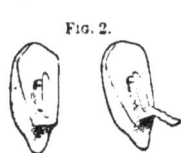

FIG. 3.   FIG. 4.

FIG. 2. will unite the strip to the
pins, and form a staple, giving
a firm hold in the rubber.

Where there is a narrow neck
of rubber running out to a single tooth,
this strip may be extended down into the
rubber to give it additional strength: in
such case the metal should be serrated at
the edges, or have small holes drilled in it.

In arranging blocks or gum teeth, the
approximal faces of the gum should be
fitted accurately, and, when practicable,
with a very thin wheel cut grooves in the
base and lower part of the approximal
faces: the rubber flowing into these grooves
materially assists the pins in retaining the
teeth in their position.

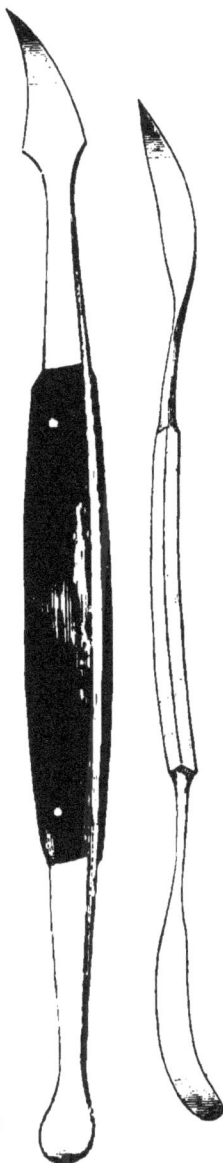

When the teeth are in proper position
upon the model plate, with a wax-knife
(Fig. 4) warmed, place wax or gutta percha
around the teeth wherever it is intended
the rubber should come, making a model
of the required rubber set. Care should be
taken not to allow melted wax to run in
between the teeth where you do not wish
the space to be occupied by rubber. In
finishing the model set, it should be
smoothed down with a warm wax-knife
(the rounded end of the one represented by Fig. 3 is very

serviceable for this purpose), and left rather thicker than the plate is required to be when finished. Just sufficient additional thickness should be given to allow for trimming and polishing: a large excess will cause a loss of material and of time in finishing.

In upper sets, when there are only a few under front-teeth remaining, and they are irregular as to position or height, it is advantageous to form an articulation, with rubber, just back of the points of the artificial teeth. This is done, after the teeth are arranged upon the model plate, by placing soft wax upon the back of the teeth, and allowing the patient to bring the jaws together sufficiently to form the desired articulation. An impression of the antagonizing surfaces of the under teeth is thereby secured, which must be carefully preserved in the model, and also in finishing the rubber set. This will in many cases prove of great utility, giving a bearing for all of the natural teeth, which could not be obtained otherwise without making the artificial teeth irregular and unsightly. Again : there are many cases, when the front teeth are properly articulated, in which we find some one or more of the inferior molars projecting up so far as not to admit of the use of an antagonizing porcelain tooth having sufficient substance to enable it to bear the force of mastication: in such cases, rubber may be substituted with good result. The proper length, and form of antagonizing face, may be obtained with wax, on the same principle as mentioned in the preceding case.

It is a better plan, after making the model set, to insert it into the mouth and prove the correctness of the articulation. Should there be any imperfection, heat such part sufficiently to make it plastic, and bring the teeth to their proper position. This having been done, place it upon the model, and it is then ready to set in the flask.

FLASKS.

A flask is a metal box in two sections, having a lid or cap, to contain the plaster in making a mould of the model set, wherein is packed the rubber to form the rubber base. Some are made with screw-bolts, to bring the sections together after the mould has been filled, and to retain it so until the case is vulcanized; others, again, have no screw-bolts, but a clamp to bring them together and hold them in position. They are made of various patterns, and many without regard to utility, merely to sell. The keys, or guiding pins connecting the base with the upper section, should be long and fitted accurately, so as not to allow any lateral motion when the empty sections are brought together. The star-flask fulfils this important indication better than any other in the market.

Fig. 5.

FILLING THE FLASK, OR MAKING THE MOULD.

Before casting into a metal flask, it should receive a coating of some substance, to prevent the plaster from adhering to it. A solution of soap, or of liquid silex, answers this purpose well. From a few trials made recently of the latter, I am induced to believe it will prove to be better than any other article used for this purpose. In cleaning the flasks after vulcanizing, when liquid silex was applied, the plaster parted readily and clean from the surface of the metal.

In filling a flask after it is prepared, saturate the model with water, to prevent it from robbing the new plaster of its moisture; mix the plaster as thick as it will pour, to give it strength; partially fill the lower section; then introduce the model, not horizontally, but slightly inclined, so that when it is forced down into its bed the air shall be excluded from beneath it. The proper height to fill the lower section, or to have the line of division, is, for teeth whose bases rest upon the base-plate, just below the porcelain (see Fig. 6), so as to allow the teeth to be imbedded in the upper section of the flask.

FIG. 6.

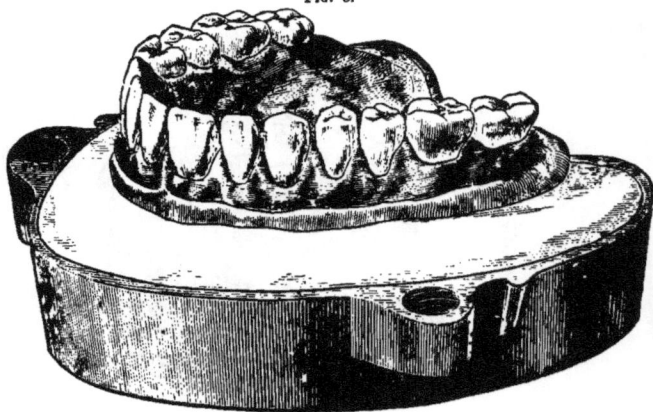

For teeth whose bases rest upon the natural gum, the division should be at the points of the teeth, thus fastening them in the lower section with their bases in contact with the model.

After the plaster has set, trim the surface smooth, and

give it a thin coating of oil or solution of soap, to prevent the upper section from adhering, being careful not to let it come in contact with the teeth. Now place the ring forming the upper section in its proper place upon the under section, using care that no plaster is interposed to prevent the forming of a good joint. Then mix the plaster as thick as it can be manipulated to make a good cast, remove all air-bubbles, and fill this ring full; place the top piece on, press down, and allow to remain until the plaster is set.

PARTING THE FLASK.

Before attempting this, it should be moderately heated. Dry or moist heat may be used. Dry heat is more uncertain in its results than moist, for the reason that dry heat expels the moisture from the plaster, and, if carried too high, the wax melts and runs into the plaster, or if gutta percha is used it becomes adherent and is very difficult to remove. When this difficulty occurs with wax, it may be partially overcome by melting it out over the fire, or, better, by immersing the flask in boiling water with the face of the mould upwards; but in either case the face of the plaster is liable to be contaminated with wax.

Moist heat is preferable; and the better plan is to apply it in this manner. Place the flask in a vessel of water, introduce beside it a thermometer, gradually raise the heat to 130° F. when prepared gutta percha is used for a model, to 120° F. when pure wax is used, or to 110° when the model is composed of wax and paraffin,—not higher; allow the flask to remain in the water at this temperature a few minutes, so that this degree of heat may penetrate through its substance; then remove and separate. When these con-

ditions are followed, the model-plate parts readily from the plaster and teeth. Should the flask accidentally get too warm, and there is wax or a wax compound present after opening, immerse it in cold water for a few minutes, to chill the surface of the wax: it will then freely part from the surface of the plaster, as it is saturated with water.

FIG. 7.   FIG. 8.

CLEANSING THE MOULD.

When the flasks are opened, the model-plate and all foreign substances should be carefully removed from the plaster mould and from the pins. A small pointed instrument, flat and curved at the point nearly at a right angle with the shaft, is very useful in cleaning under the pins (Fig. 8). It can be made out of an old excavator. It is advisable to rub the pins with a small pledget of cotton dipped in strong alcohol, especially where there are any particles of foreign substance adhering to them.

The importance of removing every particle of wax from the mould is demonstrated by some experiments made by the writer to ascertain the effect of lard-oil, yellow wax, white wax, and paraffin upon rubber during vulcanization. Rubber was vulcanized in contact with these substances separately; and in each case the integrity of the rubber was destroyed. The oil acted more energetically than the other substances, leaving the mass, when cold, semi-fluid.

The next step is to cut grooves, or gates, in the plaster, radiating from the mould to the outer edge of the flask, to allow the excess of rubber to flow out when the sections of the flask are being brought together. Fig. 9 represents the lower section so prepared.

Fig. 9.

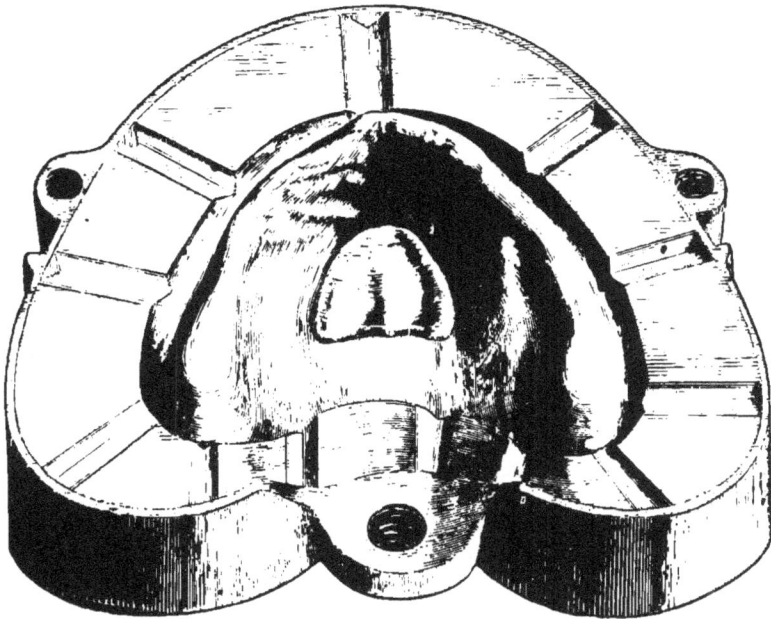

Another method is to cut a groove encircling the mould about midway between it and the edge of the flask, then making numerous small grooves radiating from the mould to the encircling groove, and from this a few larger ones to the circumference of the flask. Fig. 10 represents the upper section prepared in this manner. This having been done, the mould should be carefully brushed, to remove all loose particles of plaster.

2

FIG. 10.

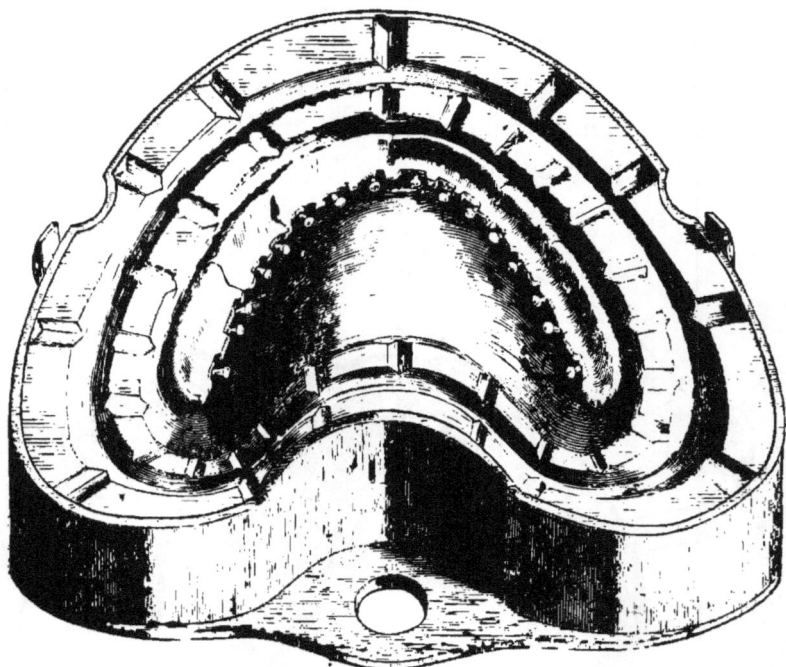

COATING THE MOULD.

The mould should then be coated with some substance, to prevent the rubber from penetrating and adhering to its surface. Either liquid silex, collodion, or tin-foil will perform this office. The writer esteems them in the order in which they are named.

Soluble glass, or liquid silex, prevents the adhesion of the rubber during vulcanization, and is readily removed from the surface in finishing; but it must be used with care. With a small brush, give the face of the mould a thin, uniform coating, allowing it to dry before the packing is commenced. This precaution is necessary to insure success; for if the liquid silex is in excess and fluid, it will insinuate itself between the pieces of rubber in packing, and prevent their adhesion.

Collodion answers the purpose well, and has been successfully used for many years. It is applied in the manner directed for using soluble glass. The only objection to it is, the film will sometimes adhere to the rubber when vulcanized, and so darken it as to give it an unsightly appearance.

TIN-FOIL.

When tin-foil is used, give the surface of the model a coating of varnish; then, while it is still adhesive, carefully press down upon it a sheet of tin-foil, making it fit accurately. After the case is vulcanized, the tin may be removed by the application of either nitric or hydrochloric acid.

FILLING JOINTS.

To fill the joints between the blocks or gum teeth, to prevent the rubber from insinuating itself, various substances have been recommended,—viz. calcined plaster colored with vermilion, finely pulverized silex or felspar moistened with liquid silex, liquid silex, and os-artificial. All of these, in course of time, will yield to the action of the fluids of the mouth; and then the ill-fitted joints will be receptacles for soft particles of food, which will be more objectionable than having them filled with good solid rubber. The best filling is an accurately fitted joint; when so made, if the enveloping plaster is of good quality and properly mixed, and no undue force is used in bringing the sections of the flask together, there is little danger of the rubber insinuating itself into the joints.

FASTENING LOOSE TEETH.

Sometimes, in opening the flask or cleaning out the mould,

a block, or oftener a single tooth, becomes detached from its bed. When such accidents occur, place a drop of thick shellac varnish upon the plaster bed, then force the block or tooth into its position: in a short time the plaster will absorb the alcohol from the shellac, and it will be firmly held in its place.

PACKING THE MOULDS.

One important consideration should ever be borne in mind to insure success in this part of the process; that is, the moulds, instruments used, and the rubber, must be perfectly clean and free from all foreign substances.

The flasks or moulds may be packed cold; but it is a much better plan to warm them. The heat should be raised gradually up to the point at which rubber is softened.

Cut the rubber in narrow strips, say half an inch long, and in small squares, and heat it to make it more plastic. The heating may be done over the naked flame of a spirit-lamp, or upon a metal plate. But by either of these modes the rubber is liable to become overheated and its texture impaired. The better plan is to use steam heat. A metal water-box, with a tube in the top for the escape of steam, may be used without danger of overheating the rubber.

Fig. 11 represents an instrument designed for the double purpose of heating the rubber for packing on the cover, and for heating the flasks in water in bringing them together after they are packed. It possesses a capacity to contain four flasks, and has the advantage over the metal water-box, in having boiling water ready in which the flasks can be immersed as soon as they are packed.

When the rubber is softened, with a small curved-pointed instrument (Fig. 7) pack the narrow strips around the pins and into the grooves and small recesses; then, with a larger instrument (Fig. 3), proceed with the filling of the mould with the square pieces, using care to consolidate the pieces as

FIG. 11.

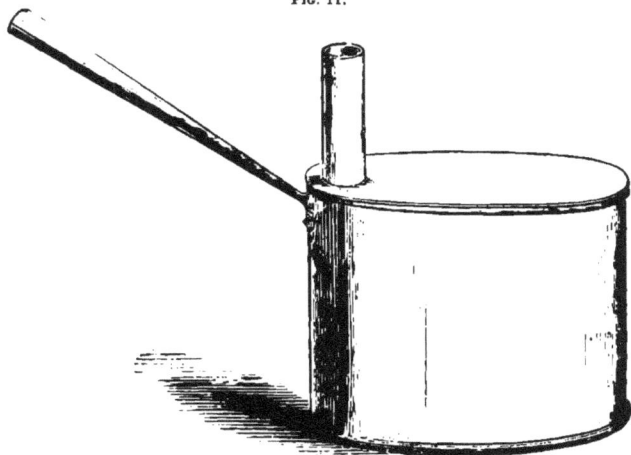

added, and also not to inclose any particles of plaster or air-bubbles.

Avoid packing full to the porcelain gums, especially when they are thin and come near the model when the flask is closed, as in such case the force in bringing the sections together would be in a great measure expended laterally against the porcelain shell and endangering fracture. But by making the centre the fullest, the force is expended upon the strongest part of the mould, and, as the rubber yields, it flows in and around the more delicate parts.

QUANTITY OF RUBBER TO FILL THE MOULD.

Rules have been given to ascertain, by measure and by weight, the exact quantity of rubber required to fill the mould.

By Measure.—Upon removing the model plate, submerge

it in water in a glass vessel, note the rise, then remove, and add rubber until the water rises again to the same point.

Fig. 12.

The above cut represents a very useful and simple instrument, invented by Mr. E. T. Starr, for measuring the quantity of rubber. With the screw set the lower point to the height of the water in the vessel; then throw in every particle of the model plate, and set the upper point to the rise of the water. The vessel may now be emptied and rinsed, to insure the absence of all foreign matter liable to contaminate the rubber, and again filled with water to the lower point; then add a sufficient quantity of rubber to cause the water to rise to the upper point, and there will be just sufficient to fill the mould.

By Weighing.—The specific gravity of wax is 0.96. I have found that of the American Hard Rubber Company's prepared gutta percha to be 2.454, and the same company's red rubber to be 1.572. Hence, to fill the mould, when pure wax is used for a model plate, it will require to one part of wax, by weight, 1.6 of the company's red rubber; and when the plate is made of prepared gutta percha, it will require to one part of it, by weight, .6 of red rubber. When the wax is colored, the disparity in weight will not be so great as with pure wax.

Of the two methods to ascertain the quantity of rubber. that by measure offers an advantage over that by weighing. in the facility with which it can be arrived at, especially when Starr's instrument is used, the calculation being based upon bulk only; whilst by weight, when the model plate is composed of more than one substance,—as it frequently is. of gutta percha, wax, and sometimes wires introduced to give stiffness,—quite an intricate calculation must be made to ascertain the exact quantity.

In filling the mould, a *small excess* of rubber should be used, to insure compactness. It is a much better plan to waste a little rubber than lose a set of teeth,—which accident is liable to occur where too close a calculation is made to save a few scraps of rubber.

BRINGING SECTIONS OF THE FLASK TOGETHER.

The moulds having been packed, the two sections of the flask are placed together and held by a clamp, or the screws inserted, whichever is used in the flask.

Although rubber is plastic, yet such is its nature that it

will not yield to the sudden application of force as well as to continued firm pressure: therefore the better plan is to apply the force through the medium of a spring clamp, which acts uniformly and continuously, which is less liable to fracture the porcelain gum than when the force is abruptly applied. Owing to simplicity in the arrangements, flasks with screw-bolts are now almost universally used.

In closing the flasks, they should be heated, to make the rubber more yielding and adhesive. Dry or moist heat may be used. In using dry heat there is danger of overheating the rubber and thereby injuring its texture. The application of moist heat is decidedly preferable. Place the flask in boiling water, frequently gently tightening the screws until the sections are brought together. This method has the advantage over the dry process in the saving of time, and of not endangering the integrity of the rubber.

Care should be exercised to bring the sections close together. If this precaution is neglected in partial sets, where the bases of the teeth do not rest upon the gums, they will be displaced from their relative position towards the natural teeth. If the teeth rest upon the gums, and the line of division is at the points, there will be a sheet of rubber overlapping them,—which will be troublesome to remove; and in a whole set there will be an extra thickness under the teeth, making them longer than is desired.

When the flask is closed, the pressures holding the sections together should not be removed until the work is vulcanized and *cold*.

VULCANIZERS.

To harden or vulcanize rubber requires a high degree of heat, and this to be maintained for a time proportioned to the temperature. As a medium, hot air, hot water, or steam may be used.   Fusible metal, wax, paraffin, and other sub-

FIG. 13.

WHITNEY'S IMPROVED VULCANIZERS.

ALCOHOL CAN.

HAYES' HIGH PRESSURE VUL-
CANIZING OVEN; ALSO NEW VUL-
CANIZING BOILER.

stances capable of sustaining without change the required degree of heat, have been resorted to. Either water or steam confined in a steam-tight vessel, called a vulcanizer, is used by dentists. A vulcanizer with a single chamber, which is but a modification of Papin's digester, is now universally used. The usual practice is to vulcanize with the flask covered with water; but if the operator desires to vulcanize in steam, using a vulcanizer with a single chamber, it can be readily done by placing a diaphragm above the water and allowing the flask to repose upon it.

Every vulcanizer should have a safety-valve, or its equivalent; also, a *correct* thermometer, or steam-gauge, to enable the operator at all times to know the extent of pressure within.

A vulcanizer should be strong enough to resist far greater pressure than is usually employed,—for two reasons: first, it is continually becoming weaker every time it is used, from the strain upon the fibres of the metal and erosion of its substance; (this latter cause of weakness has been disputed; but I have repeatedly detected copper in the dark deposit which forms upon the inside of the vulcanizer,—which proves a loss of metal, and, consequently, of strength;) and, second, to avoid the danger of an explosion in case, by inadvertence, the pressure should become unduly great.

Wrought is preferable to brittle cast metal for a vulcanizer, as it has greater tensile strength, and in case of an explosion it generally opens in a rent, while cast metal separates into fragments,—which, when propelled by a force of ninety or a hundred pounds to the square inch, would be rather unpleasant companions.

ELASTIC FORCE OF STEAM.

As high steam is used in vulcanizing, it is important that

the operator should be conversant with the nature of the agent which he employs to accomplish his end. It is perfectly safe; but the following will show him that it must be used with discretion and judgment. Numerous experiments have been made by scientific men to ascertain the elastic force of steam at different temperatures. The results of their investigations are not uniform; although they all agree in showing the immense force exerted by this agent at high temperatures. Haswell's tables are looked upon as good authority. The results of the investigations of the Franklin Institute Committee, in the higher degrees, give a greater elastic force than the table below quoted. I shall, however, quote the results of the experiments of the commission of the French Academy appointed by the French Government to investigate this subject, for the reasons that, from the manner in which they were conducted, they are probably as reliable as any, and that they are extended to a more elevated temperature than the others.

Elasticity of steam, taking atmospheric pressure as unity.	Tempera-ture F.	Pressure per square inch, lbs.	Elasticity of steam, taking atmospheric pressure as unity.	Tempera-ture F.	Pressure per square inch, lbs.
1	212°	14.7	8	341.78°	117.6
1¼	233.96°	22.05	9	350.78°	132.3
2	250.52°	29.4	10	358.88°	147
2½	263.84°	36.75	11	366.85°	161.7
3	275.18°	44.1	12	374.00°	176.4
3½	285.08°	51.45	13	380.66°	191.1
4	293.72°	58.8	14	386.94°	205.8
4½	300.28°	66.15	15	392.86°	220.5
5	307.05°	73.5	16	398.48°	235.2
5½	314.24°	80.85	17	403.82°	249.9
6	320.36°	88.2	18	408.92°	264.6
6½	326.26°	95.55	19	413.78°	279.3
7	331.70°	102.9	20	418.46°	294
7½	336.86°	110.85			

I would here call the attention of those using high steam to an important consideration. In raising steam, *the ratio*

*of increase of pressure or elastic force is far greater than that
of the increase of temperature.*

By referring to the above table, commencing at 212° and
taking steps as near fifty degrees as is given in the ascend-
ing scale, we find this exemplified.   Thus,—

		Increase of temperature.	Increase of force per square inch.	Giving a force per square inch.
From 212°	to 263.84° =	51.84°.........	22.05 lbs.........	36.75 lbs.
" 363.84°	to 314.24° =	50.40°.........	44.10 lbs.........	80.85 lbs.
" 314.24°	to 366.85° =	52.61°.........	80.85 lbs.........	161.85 lbs.
" 366.85°	to 418.46° =	51.61°.........	132.15 lbs.........	294 lbs.

This comparison shows clearly how rapidly the pressure
increases at high temperatures, and warns the operator that
a strong instrument, combined with care and judgment in
its treatment, are indispensable to safety.   Besides the rapid
increase of pressure, it must be borne in mind that, at high
temperatures, copper, of which the boiler is composed, becomes weakened, and in a measure loses its power to resist
this great imprisoned force.   Copper in passing from 212°
to 320° F. loses about one-tenth of its strength, and at
550° it has lost one-fourth of its tenacity.

VULCANIZING.

It is a better plan for the novice, or when first using a
new vulcanizer, or after setting a new thermometer, to put
in a test-piece, in order to ascertain to a certainty if the
work is properly done, before the flask is opened.   This
may be done by enclosing a small piece of rubber in plaster,
then enveloping the whole in tin-foil to prevent the water
from disintegrating the plaster, and placing it in the vulcanizer beside the flask.   When the vulcanizer is opened, take
out the test-piece, and cool it: if properly done, a thin strip
is somewhat elastic, but so hard that when it breaks it is

with a clean fracture, cuts tough, and gives a shaving. on scraping, like that of horn. If overdone, it is dark. brittle, and the shavings will not curl up, but are short and brittle. If underdone, it is not elastic, but will remain as bent, and cuts like a piece of leather. In this case the vulcanizer must be closed up, and the process continued for a time proportioned to the condition of the test-piece. It must be borne in mind, in examining the test, that rubber, when properly hardened, when heated above the boiling-point of water is quite flexible.

Place the flask—or flasks, if more than one case is to be vulcanized at the same time—in the vulcanizer; add water to cover the flask at least an inch; dust the top of the packing with pulverized soapstone, or whiting, to prevent the cap from adhering; then place the cap on and screw down firmly. When the heat has gone up to about 240°, tighten the nuts again, as packing yields when heated, especially if new, and without this precaution might cause a leakage of steam.

The next step is to apply the heat. Gas, alcohol, and coal oil are used for fuel. Most vulcanizers are constructed to burn either gas or alcohol, as may best suit the convenience of the operator. Regulate the flow so that the flame shall not deposit carbon on the bottom of the vulcanizer: this deposit, when formed, acts as a non-conductor, and in a great measure neutralizes the effect of the flame. Raise the heat gradually up to 320° F. The time expended in raising the heat should not be less than half an hour, and when there is a thick mass of rubber the time should be extended to even an hour. When the heat is forced up too rapidly, it frequently causes the rubber to be porous or spongy.

When the vulcanizing-point (320° F.) is attained, lower the flame so that the heat shall remain stationary until the end of the operation. To maintain this point requires care and watchfulness on the part of the operator; and, after the remarks upon the elastic force of steam, I trust he will fully appreciate the danger of negligence.

The time the heat is to be held at 320° F. to vulcanize varies, in different makes of vulcanizers,—probably owing to want of uniformity of the thermometers used,—from one hour to an hour and a half; also, different varieties of rubber require different lengths of time to vulcanize. A good plan is to test a new vulcanizer with a trial-piece, before attempting to vulcanize a set of teeth, taking for time one hour and a quarter and noting the effect.

Rubber may be vulcanized at a much lower heat than 320°; but the time must be proportionally extended. A higher temperature may be used,—say 330°,—and the time shortened; but it is not advisable to carry the heat beyond 320°. When the heat is carried very high, it causes the rubber to darken and become brittle, and the toughness and elasticity so essential for dental purposes are lost.

The preceding remarks apply to the use of the red rubber, which is composed of caoutchouc, sulphur, and vermilion. When the brown (or black rubber, as it is sometimes styled), composed of simply caoutchouc and sulphur (caoutchouc 2, sulphur 1), is used, great care should be exercised in raising the heat, especially in thick pieces. When at 320° F. it will vulcanize in the same time as the red. This kind of rubber, although unsightly in appearance, is growing in favor for dental purposes, owing to its possessing greater strength and toughness than the highly-colored mixtures. In using the light-pink rubbers, composed of caoutchouc, sulphur, vermilion, and white clay, or oxide of zinc, owing to their containing so much foreign matter (frequently forty-eight per cent. of clay), the heat may be more quickly raised with safety; and when at 320° F., the time to be held at this point to vulcanize is reduced to three-fourths or one-half of that for the red rubber.

To insure success and produce the best results in hardening any kind of rubber, the heat should be *gradually* raised to the vulcanizing point, not higher than 320° F., because the best quality of rubber may be rendered worthless by the quick process and vulcanizing at a high range of temperature.

When the time has expired, cut off the flame and allow the vulcanizer to cool down to 212°. This may be hastened by letting off the steam; but it is much better to allow it to cool gradually without doing so. Now loosen the nuts, and take off the cap or head. If a trial-piece is used, take it out and examine. If properly done, remove the flask, and allow it to cool gradually. It must not be placed in cold water to hasten the cooling, as this would endanger fracturing the porcelain.

OPENING THE FLASK.

When the flask is *cold*, remove the top piece, then carefully pry the sections apart, commencing by insinuating the point of a knife between the joints, at different points, until it yields. Then, with a pointed knife, cut away the plaster near the margin of the flask, until the central part containing the case may be removed; or, where the flasks have much bevel, by gentle taps with a hammer upon the metal the whole of the plaster will separate in a body. The plaster readily separates from the rubber, if the mould has been coated as directed. Now wash, using a stiff brush, and the case will be ready for the finishing process.

At this step the operator should immediately remove all the plaster from the flasks, and wash and dry them. When first opened, they are much more easily cleaned than when the plaster has been allowed to dry and become cemented to the metal by oxidation: besides economy in time, they are in a good condition when next required for use, and will last much longer than if left with the plaster adhering. The vulcanizer should also be washed, to remove the deposit formed within it.

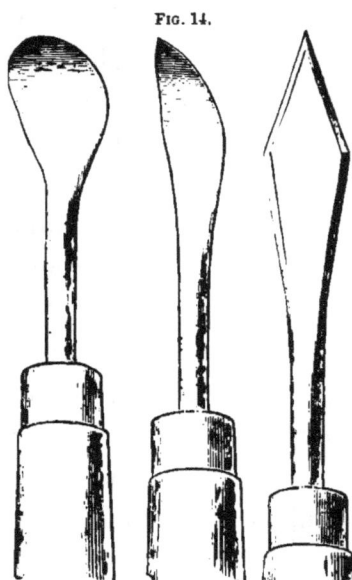

FIG. 14.

FINISHING.

With coarse files remove the surplus rubber. Files are made especially for this work, straight, half-round for the outer edge, and curved for the lingual surface of the plate. Scrapers then come in

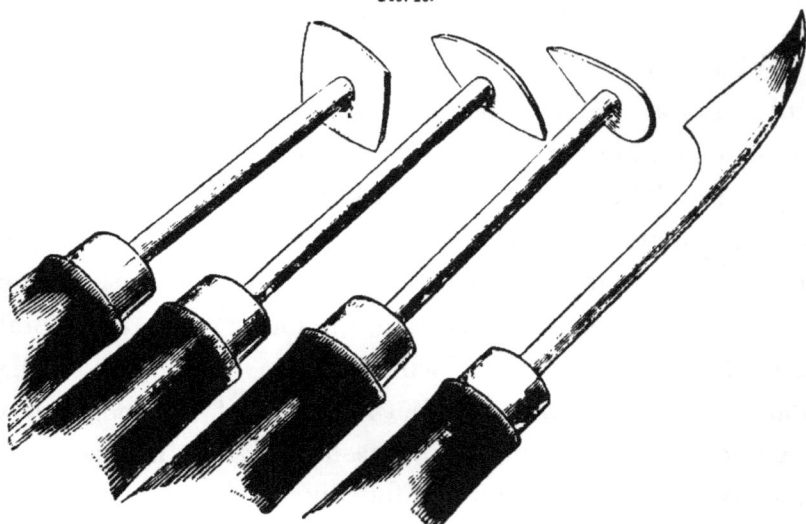

FIG. 15.

play, to remove such surplus as is not readily reached with the file, and also to obliterate the file-marks. The curved form is very useful on the lingual surface. With a graver cut away the excess from around the teeth, making the joining even and smooth. Coarse burrs for the lathe, made for this purpose, cut more rapidly than the file or scraper; but, without great care, there is danger of cutting through the plate: on this account, their use is not recommended. As a safeguard against the unpleasant accident of cutting through the plate in dressing it down, from time to time use the callipers to ascertain its thickness. When reduced to the proper

FIG. 16.

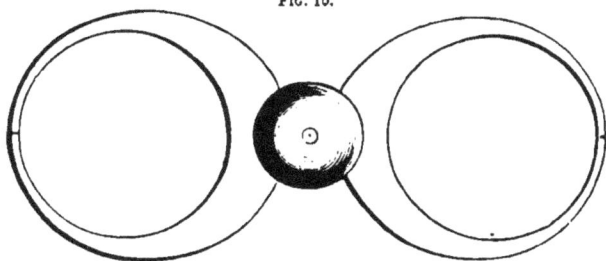

FIG. 17.

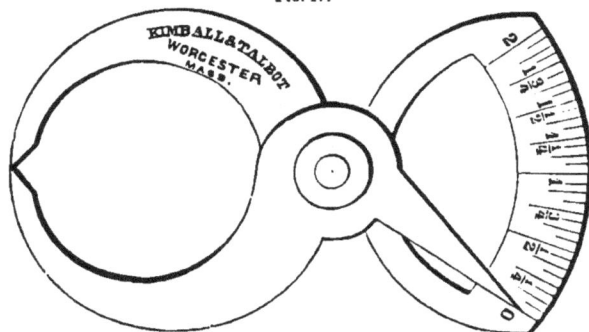

thickness, smooth with fine sand-paper. Then, to prepare for the polish, use very finely pulverized pumice-stone, made into a paste with water. It may be applied by using a stick of soft, porous wood: cottonwood is the best for this purpose.

as it is very porous and tough.   Or it may be applied on a cork wheel in the lathe ; or, better, with a felt wheel.

When all the scratches have been obliterated, proceed to polish.   This may be done by using a cotton buffer, or a very soft brush wheel, on the lathe, with calcined buck-horn, or prepared chalk, free from grit, moistened with water.   In giving the finishing touches with the polishing material, have it diluted very thin, and the wheel running at high speed, at the same time giving the work a vibrating motion. To give an exquisite finish, then apply fine rotten-stone, free from grit, mixed with olive oil on chamois-skin, or on the hand.   Remove the oily coating with dry rotten-stone, magnesia, or fine zinc white.   The burnisher may be used with advantage on parts not otherwise accessible.

To insure success, every stage in the process of finishing should be complete in itself, and the work should be washed before proceeding to the next.

The palatal surface of the plate cannot be dressed down and polished without destroying the accuracy of its adaptation.   Hence the necessity of having the face of the model upon which it reposes when in the plastic condition, smooth and free from imperfections.

After the use of the files and scrapers, very finely pulverized silex may be substituted for sand-paper and pumice. Many prefer this mode of procedure.   Scotch-stone in slips is also used for the same purpose, and is very effective : it cuts fast, and leaves a smooth surface.

PARTIAL SETS.

The general instructions for the getting up of whole dentures will apply to the mode of procedure in making partial cases.

The articulation is made in the same manner as for metal work. Where there is a narrow neck of rubber running out from the plate to connect with a single tooth, give it additional thickness, to render it sufficiently strong to resist the force that may be applied to it. After the model plate is made, place it in the mouth, and correct defects in articulation, if there be any.

If the base of the teeth rest upon the plate, and not upon the gums, now cut off the plaster teeth on the model, leaving about one-sixteenth of an inch remaining, to mark in the rubber their position when finishing the plate. In this case the porcelain is attached to the upper section of the mould. If the plaster teeth were left remaining, they would be fractured on separating the flask, and might injure the model. This having been done, place the model set on the model, and proceed to make the mould as for a whole set. But if the base of the teeth rest upon the gum, place the model set on the model, and, in filling the under section of the flask, bring the plaster up to the points of the teeth, to hold them fast to the model, making the line of separation of the sections of the mould at that point. Now trim off such portions of the plaster teeth as may be liable to be fractured or which would interfere with the free parting of the mould. Then make the upper section.

Clasps.—Preference, in all cases where admissible, should be given to atmospheric pressure to retain the plate in its position in the mouth; but when clasps have to be resorted to, they may be made either of rubber or of metal. When made of rubber, they are moulded with the model plate.

When made of metal, it must be either gold, or gold alloyed with platina. Silver will not answer, as rubber will not vulcanize hard in contact with this metal.

First bend the clasp to fit the tooth accurately; then make the attachment by which it is to be held to the rubber (this may be done by soldering a thin plate of gold or platina to the clasp in such a position that it will be enclosed in the

FIG. 18.

rubber); then perforate the plate with numerous small holes, which should be countersunk on both sides. This plate entering the base, the rubber filling the holes forms pins which rivet the clasp securely to the rubber plate.

Or the attachment may be made in this manner. On the parts of the clasp that can be covered with rubber, drill one, two, or three holes, as the space may admit; insert gold or platina wire, solder with gold solder, then cut off at

FIG. 19.

proper length, and head them; these act in retaining the clasp in the same manner as the double-headed pins in securing the tooth to the base, and offer the advantage over the perforated plate of being more easily manipulated, and less liable to become displaced in packing the mould. The clasp is to be attached to the model plate, and will remain secured in the mould when it is opened.

ATTACHING TEETH TO A GOLD OR PLATINA PLATE BY MEANS OF RUBBER.

Single teeth, plain or with gums, and blocks, may be secured to a gold or platina plate with rubber, enabling the operator in many cases to build up and restore the parts lost by absorption, and to make more accurate joining than he could otherwise do in using a metal base-plate. The method for doing this has been given to the public by Dr. Wm. Hopkinson, published in the "London Dental Review" of October and in the "Dental Cosmos" of December, 1860. His instructions are so explicit that I shall copy them, with slight modifications:—

"A plate having been struck up, the teeth are prepared and mounted upon it with wax, and the bite properly adjusted in the usual manner when vulcanite is used. A pointed broach is to be passed between the teeth and wax at intervals, so as to mark the plate, keeping as near the teeth as possible. The teeth and wax are then to be removed, and holes are to be drilled corresponding to the marks on the plate. Into these holes gold" or platina "wire" of the proper size to give the "strength required is to be soldered, and the ends of these wires are then to be bent round in the form of a loop, so as to touch the plate."

"A sufficient number of pins" or loops "having been inserted into the plate to make the case firm, the wax and teeth must be replaced as at first. The wax should be shaped on the lingual side of the teeth, taking care to cover the looped pins, so that they may remain out of sight when the case is finished." In setting single plain teeth, "if it is necessary to build out so as to represent the natural gum on the labial side," or when crossing a deep depression formed by irregular absorption, "build up with wax the same as you would in making a case in vulcanite, bearing in mind that the curved pins already soldered to the plate are sufficiently strong for the rubber in front. Having got the teeth in position, and the wax trimmed to represent the case when finished, place the model with the piece in the vulcanite flask, and proceed in the usual way for vulcanizing."

The writer has inserted single teeth on gold plate upon this principle, by soldering gold stays in the parts concealed by the rubber, and then perforating them full of small holes. This work has stood the test of three years' wear without signs of deterioration.

REFITTING GOLD PLATES BY MEANS OF RUBBER.

A temporary or other gold plate which in consequence of absorption of the alveolar ridge can be no longer worn may be made to fit accurately by adding a plate of rubber upon its palatal surface.

An excellent method is described by Professor Richardson, in the " Dental Register of the West" of March and the " Dental Cosmos" of April, 1861, which, with some modifications, is repeated below:—

" Perforate the palatal portion of the plate with from eight to twelve holes at different points, and also the external borders from heel to heel of the plate, at intervals of from one-eighth to half an inch apart, and near the edges. These holes may be enlarged to the dimensions of a medium-sized knitting-needle." On the lingual and buccal surfaces the holes are well countersunk with a burr drill. " Employ this plate as a cup or holder; take an impression of the mouth in plaster, pressing the plate up closely to the parts," and at the same time preserving the articulation. " The plaster forced through the holes, and filling the counter-sinks on the opposite side of the plate, will serve to bind the plaster to the plate, and prevent, with cautious manipu-lation, the two from separating as they are being detached from the mouth." "When removed, the plaster impression lining the plate is trimmed even with the borders of the latter," and coated with a solution of soap. "The lower section of a vulcanizing flask is now filled with a batter of plaster on a level with its upper surface, and the impression, filled with the same, is turned over and placed in the centre of the flask, with the edges of the plate touching the surface of the plaster." Now trim the plaster, so as to free the plate that it may part with the upper section of the flask. Remove the plaster from the holes and countersinks, and fill with wax. Give the plaster a thin coating of oil or of a solution of soap; the wax in the countersinks should also have a very slight coating of oil; adjust the upper ring of the flask, and "pour the plaster in upon the upper surface of the plate and teeth, filling it;" then place the cap, or upper piece on.

When the plaster has become hard, carefully separate the sections, remove the plaster impression from whichever section it may adhere to, and also, with a small instrument, all particles of wax from the holes and countersinks. We then have a perfect model in the lower section, and a mould of the space necessary to be filled to restore the proper adap-tation of the plate. Cut grooves or gates running from the edge of the mould to the edge of the flask. Coat the surface of the model, as before directed, to prevent the adhesion of the rubber. "A sufficient quantity of vulcanizable rubber is now placed upon the model or packed upon the palatal surface of the plate." "The whole being sufficiently heated.

the two sections of the flask are forced together, expelling the redundant material.   The piece is then vulcanized." "The union between the vulcanite lining and the plate will be strong and lasting, and altogether impermeable to the fluids of the mouth."   "In lower pieces the holes should be made along the external and internal borders."

Refitting Vulcanite Sets.—In doing this, the process is precisely the same as for a gold plate, with the exception that the holes may be "made twice or three times the size," and the palatal surface should be scraped perfectly clean before the new rubber is added.

TO MAKE A NEW RUBBER PLATE AND PRESERVE THE ARTICULATION.

In the preceding article, the method is given to refit a rubber plate when its adaptation was lost by absorption; but when it is desired to form an entire new plate and preserve the articulation, a different mode of procedure is required.

Roughen the palatal surface of the rubber plate, to cause the plaster to adhere to it; then use it as an impression-cup to take a plaster impression, being careful when it is in the mouth to preserve the articulation.   In this impression cast the model; trim; cut keys or conical holes at several points in its outer face.   Now, before separating the impression from the model, make a cast of the face of the teeth in two or three perpendicular sections, extending to the base of the model, using a solution of soap or other parting substance on the plaster mould.   Remove this mould of the face of the teeth, which indicates their true position relative to the model; then take the impression from the model.   By the aid of heat sufficient to soften the rubber, remove the teeth from it.   Next make a model plate with prepared gutta percha.   Now secure the sections of the mould of the face of the teeth to the model (their place will be indicated by the keys), adjust the teeth in their proper positions in the plaster mould of them, and build up with gutta percha or wax to the proper form of the model set.   This being done, test its accuracy of contour and articulation by placing it in .the mouth.   Then, using the model, proceed as for making a new set.

To Improve the Color of Rubber.—After the case is finished, place it in a clear glass vessel, cover it with alcohol, and expose to the action of the rays of the sun from six to twelve hours, according to the change it is desired to make in the color.

By this process the red rubber is brought to a bright red. Pink and other light rubbers require this treatment to develop their colors.

Alcohol, after having been used several times, loses its bleaching effect.   The vessel containing the case should be covered with a piece of glass, to prevent loss of alcohol by evaporation.

Repairing.—If a tooth or a block has been broken, remove the remainder, and cut an irregular, dovetailed groove in the base, occupying the space to be supplied with new rubber.   Arrange the tooth or block in its proper position.   Fill the groove with wax, giving a little more fulness than the surrounding surface, to allow for finishing.   Paste a strip of thin paper on the front of the block, allowing it to extend over its base and the adjoining teeth, to prevent the plaster from insinuating itself into any space under the base that may not be occupied by the wax.

Now fill the lower section of the flask with plaster, also the palatal surface of the denture; turn it over, and press it down in the centre of the flask. The teeth, and all of the parts except the wax and its immediate surroundings, are to be imbedded in the plaster of the lower section. Trim, using precautions to prevent adhesions, and fill the upper section. After the plaster has become hard, the flask may be opened without the aid of heat. Now remove every particle of wax, and cleanse the cavity perfectly. A pledget of cotton dipped in absolute alcohol may be used with advantage to effect this end: without this precaution the newly-added rubber will not unite with the base. Cut grooves, or gates, pack the mould, bring the flask together, and vulcanize in the same manner as for a new set, giving the same heat and time. The extra heat employed in the second vulcanizing renders the parts previously hardened much darker. The color may be restored by wetting the surface with dilute nitric acid, washing with water, and then immersing the piece in an alkaline solution to remove all traces of acid. This speedily restores the color; but the use of nitric acid is objectionable, as it appears to injure the texture of the surface of the rubber. A better plan, though not so prompt, is to bleach it in alcohol.

To Bend Hardened Rubber.—Hard rubber is readily bent when heated to a proper degree, and, after cooling, retains its shape. From numerous experiments, its greatest flexibility appears to be in the range of temperature of from

240° to 280° Fahrenheit; above 280° it loses its tenacity in proportion as the temperature is increased.

Small pieces of uniform thickness, such as a clasp, may be softened so as to be bent, by oiling and holding over the flame of a spirit-lamp; but where the piece is of uneven thickness the thin parts are liable to become overheated before the thicker portions are rendered pliable.

Where the piece is large, or of irregular thickness, the better plan is to immerse it in a vessel of oil, and raise the heat to the proper degree, which may be tested by introducing strips of rubber: by this means the whole piece becomes uniformly heated. Sometimes by this method a misfit may be corrected by making a correct model and pressing down upon it the softened plate and retaining in place until cold, using the precaution of protecting the hands by interposing a napkin between them and the heated surface.

To Set a New Thermometer.—Place a little white or red lead, ground in oil, on the threads of the screw; then screw it firmly into the cap. The lead assists in making a steam-tight joint.

Packing—is made with a web in its substance, and also without it. The kind having no web in it is not reliable, as it becomes so soft and yielding when exposed to the heat and pressure required for vulcanizing that if the screws are not frequently tightened during the operation it is liable to be blown out, or the joint opens, causing a loss of steam. That containing a web should be selected. In using this it is necessary to tighten the screws when the heat comes up to about 250°, for two or three times after it is first put in, after which it will not yield.

When new packing is required, first remove all of the old, and, if the bolts securing the cap pass through the packing, place the cap upon it, and, with a pointed instrument, through the bolt-holes mark their size and position, then cut out those parts, and place the packing on. Before the cap is screwed down, dust the top of the packing with whiting, or, better, pulverized soapstone, so that it be well covered: this will prevent it from adhering to the cap and tearing in opening the vulcanizer.

Keep the pulverized soapstone in a broad-mouthed vial, having the orifice covered with thin gauze. By dusting the soapstone on in this manner through the gauze, it is spread evenly.

To Separate Teeth or Blocks from Vulcanite Base.—The rubber must be softened by heat. The heat may be applied by holding the case over the flame of a lamp, upon a metal plate or shovel over the fire, or immersing it in heated oil. I give the preference to the latter method, as the whole substance is more uniformly heated. When the rubber is softened, insert the point of an instrument between it and the porcelain, which will cause the tooth to separate readily. Where the pins are headed, the rubber generally adheres under the heads. This should be removed while the tooth is hot.

To make a Solution of Soap for Parting Plaster.—Use white Castile soap. An ounce of soap to a pint of water is a good proportion. Cut the soap in thin shavings, put it into water, and raise to the boiling-point. When the soap is dissolved, bottle it.

When using the solution, pour out just enough for present use. Do not place the brush in the bottle after using it upon a plaster surface, as this would cause the solution to become turbid. The same effect would be produced by pouring the residue, after using, back into the bottle. I much prefer a solution of soap to any other substance used in parting plaster.

Coloring Plaster for Impressions.—Sometimes, through some inadvertence of the operator, the impression will adhere to some part of the model. In such a case, it is of great advantage to have the two parts of different colors, so as to enable him to detect the line of separation.

To color plaster for the impression, add to the dry plaster a very small quantity of Venetian red, or, better, vermilion: this, when mixed with water, will give it a pink color. A small quantity of coloring-matter does not injure the texture of the plaster, and renders it more pleasing to the eye.

To Hasten the Hardening of Plaster for Impressions.—This may be done by either mixing the plaster thick, adding to it common salt, or by using hot water. The hardening can be still further hastened by combining the three methods. When salt is introduced in the plaster, to insure a smooth surface of the model, it should be cast immediately after the impression has been taken. There are other methods of accelerating the hardening of plaster; but the above accomplishes the end without being unpleasant to the patient.

ADDENDUM.

RUBBER COMPOUNDS.

In treating upon this subject, I will premise, by stating the sole object of these investigations has been to throw some light upon the composition of hard rubber, which is now so universally used by dentists, and of which so little has been really known.

In taking up this subject, I shall first lay before the reader such information upon it, as is detailed in a few of the most interesting of the numerous rubber patents. Then relate some experiments upon the rubber compounds in the market for dental purposes, and also give the results of my own attempts to form mixtures to make hard rubber of different colors.

To form a compound of caoutchouc that will vulcanize, it is essential that it shall contain sulphur, either alone, or in combination with some other substance, as in sulphides, &c.

The earliest information we have upon this subject is found in the specification of Charles Goodyear, dated January 30th, 1844, for making soft or flexible rubber, that will resist the action of the usual solvents of caoutchouc, and will not be affected by cold or by heat, if the temperature is not raised above the vulcanizing point.

The mixture he prefers is

<table>
<tr><td>Caoutchouc</td><td>25 parts.</td></tr>
<tr><td>Sulphur</td><td>5 "</td></tr>
<tr><td>White Lead</td><td>7 "</td></tr>
</table>

"The caoutchouc is to be dissolved in spirits of turpentine, or other essential oil; and the white lead and sulphur are ground in spirits of turpentine in the ordinary way of grinding paint. These articles thus prepared, may, when it is intended to form a sheet by itself, be evenly spread upon any smooth surface, or upon glazed cloth, from which it may be readily separated."

Or, "instead of dissolving the India-rubber in the manner above set forth, the sulphur and white lead, prepared by grinding as above directed, may be incorporated with the substance of the India-rubber by the aid of heated

46

cylinders or calender rollers, by which it may be brought into sheets of any required thickness," or it may be applied so as to adhere to the surface of cloth or leather.

"To destroy the odor of sulphur in fabrics thus prepared, the surface is washed with a solution of potash, or with vinegar, or a small portion of an essential oil, or other solvent of sulphur."

"When the solvent has evaporated, the compound is subjected to the action of a high degree of temperature, which will admit of considerable variation, say, from 212° to 350° F., but for the best effect, approaching as nearly as may be 270° F. If the exposure be to a temperature exceeding 270°, it must continue but a very brief period."

In forming hard rubber compounds, to the base, caoutchouc and sulphur (or sulphur compounds) is frequently added earths, metallic oxides, shellac, resin, bitumen, saw-dust, charcoal, ground pottery, &c., for the purpose of utility or economy for articles of commerce, and some of the rubbers vended for dental purposes are so loaded with earths, or metallic oxides, as to render them unfit for the use they are ostensibly designed.

It may be interesting to refer to the original specification of the patent of Nelson Goodyear, dated May 6th, 1851, for making hard rubber. He says in treating caoutchouc for this purpose, it is combined with sulphur, the best proportion being about equal parts by weight of each ingredient. By combining sulphur in this proportion with caoutchouc, and subjecting the compound to the curing operation, a hard substance will be produced. But a still better result will be obtained by the introduction of magnesia or lime, or a carbonate or sulphate of magnesia, or a carbonate or sulphate of lime, or calcined French chalk, or other magnesian earth into the compound; in which case the following proportions will be found highly advantageous, viz.:

1 ℔ Caoutchouc,
½ ℔ Sulphur,
½ ℔ Magnesia or lime, or the carbonate or sulphate of magnesia, or the carbonate or sulphate of lime, or French chalk, or other magnesian earth.

The proportions specified in both of these compounds may be considerably varied, without materially changing the result, but in no case is it desirable to use a much less quantity of sulphur than four ounces to every pound of caoutchouc.

With either of these compounds just described, gum lac, or gum shellac, may be combined to great advantage, say in proportion of from four to eight ounces, to every pound of caoutchouc. Rosin, oxides, or salts of lead or zinc of all colors and other similar substances, both mineral and vegetable, may be added in small quantities to either of the compounds for the purpose of imparting a polish or a suitable color thereto, and for making the mixture work more easily, but no precise rule for these additions can be given; nor, indeed, is it necessary, as the taste and judgment of the operator will be his guide in this particular.

The compounds may be mixed by a masticating machine (or by any other means employed in the manufacture of India-rubber compounds) until the several ingredients are thoroughly incorporated. The mineral ingredients mixed with the caoutchouc should be finely divided, and good results are produced by reducing them before mixture to an impalpable powder.

When mixed, the compounds are either rolled into sheets, by means of calendering rollers or moulded into the desired shape.

When thus rolled or moulded, the compounds are then "cured." This is effected by exposing the compound to a high degree of artificial heat, using for this purpose either *steam, hot water or hot air.* The degree of heat to which the compound is to be exposed, and the duration of its exposure, will depend somewhat upon the size and thickness of the article; but in ordinary cases, the heat should be raised to about 260° or 270° F., and the compound exposed to such heat for about four hours; as a general rule, however, it may be stated that the heat should range from 250° to 300° F., and the time of exposure from two to six hours. The compound by undergoing this heating or curing operation, will become of a hard, stiff character, in many respects resembling tortoise-shell, horn, bone, ivory and jet.

The first record we find of the mixture of rubber for dental purposes, is in the specification of the Patent of Charles Goodyear, Jr., "for improvement in plates for artificial teeth," dated March 4th, 1855. He says, "The best compound I believe to be *one pound of* India-rubber or gutta percha, (or of the two combined in suitable proportions) with *half a pound of sulphur,* together with a suitable quantity of coloring matter.

"To obtain a suitable color, I mix with the caoutchouc or gutta percha, *vermillion, oxides of zinc,* or of *iron,* or any coloring substance that will stand the necessary degree of heat with the action of the sulphur. This compound, after having been moulded, is to be subjected to heat for about six hours, and in doing so, I gradually raise the heat up to about 230° F., say in half an hour, and then, unless there be considerable quantity of foreign matter present, the heat may be raised, quickly as may be, to about 295°, otherwise, I raise the heat more slowly, and retain the compound at about that temperature for the remainder of the six hours, and then allow the whole to cool down, when the process will be completed."

A specification bearing date of March 20th, 1861, for making pink rubber, "adapted especially for dental purposes," is of some interest, will be given in full as copied from the English Government Patent Records, viz.:

"For the purpose of this invention, I treat the red caoutchouc of commerce, and in the following manner:

"1st. I soften the red caoutchouc by dissolving it in sulphuret of carbon, ether, chloroform, or other solvent. And according to the degree of solution to be obtained, the caoutchouc must be dissolved in a quantity of the solvent varying from about one-quarter of the weight up to an equivalent weight of caoutchouc.

" 2d. I introduce into the solution one of the following substances, *sulphate of barytes*, of *manganese*, of *strontian*, of *antimony*, *calcined alumina*, *calcined or precipitated silex*, *phosphate of lime*, or *carbonate of baryta*. These agents are used alone or mixed with *oxide of zinc*.

" The object of their introduction is to reduce or tone down the red in the caoutchouc, and to cause it to assume a flesh-colored tint; sometimes a little carmine is added. The decoloring agent is introduced in about equal quantity by weight to that of the red caoutchouc. The product obtained by the treatment is in the form of a flesh-colored paste of a consistency suitable for moulding of dental pieces and other articles to be manufactured.

" When the paste is required to be more supple and malleable, a small quantity of ordinary caoutchouc in solution is added.

" 3d. The dental piece or other article having been moulded from the paste prepared as above described requires to be solidified. This operation takes place in a closed vessel, where the temperature is raised to from 300° to about 335° F. After an exposure of about half an hour the caoutchouc will be found hard, and at the expiration of three-quarters of an hour, it will be found to have acquired all the hardness necessary.

" Caoutchouc prepared in the ordinary manner would require to obtain an equal degree of hardness, at least one hour and a quarter.

" The flesh color will be improved by exposing the piece or other article either directly or in an alcoholic bath to solar rays for a period varying with the intensity of the light."

II. H. Day in his specification of patent for improvement in preparing and Vulcanizing India-rubber and Gutta Percha, dated June, 1857, asserts, by his process, very thick pieces of rubber may be vulcanized uniformly hard and solid throughout their entire substance. To accomplish this object, he mixes, " with the matter, when prepared for being vulcanized, a substance that will prevent the spongy or cellular character, by absorbing the sulphur gases as fast as generated. The material which is proposed to be employed for effecting this object is by preference ordinary pipe clay (alumine), but other substances capable of absorbing the gas may be employed.

" The gum may be prepared for vulcanizing in the following manner:

" One pound of purified caoutchouc or gutta percha having been mixed with eight ounces of sublimed sulphur in the usual manner, eight ounces of alumine are added to the mass, taking care to have it distributed evenly throughout. The mass may now be vulcanized in from four to seven hours at a temperature from 230° to 300° F., in the ordinary manner."

He further says, "Articles of great thickness, and requiring to be made hard and equally solid throughout, can thus be produced with facility, inasmuch as all the gases evolved by the sulphur will be rapidly absorbed by the alumine, or the equivalent absorbing agent employed. In this manner, balls four inches in diameter, which do not expand perceptibly when taken out of the mould, may be made, and when cut will be found uniformly dense and compact throughout."

Two different mixtures made from this formula produced a dark-colored rubber, hard and compact, but not so tough as when caoutchouc and sulphur alone were used.

In Austin G. Day's specification, we find some interesting remarks upon the nature of rubber compounds. He claims his compound to be in contradistinction from Nelson Goodyear's hard and inflexible substance, a hard but highly elastic gum compound, obtained by a process differing in the length of time, in the degree of heat and in the proportion of the ingredients and the mode of equalizing the temperature from that of N. Goodyear's.

Day's composition is simply caoutchouc and sulphur, as follows:

> For 1 ℔ of purified Para rubber...............8 oz. of sulphur.
> "   E. I. or African......................8 to 10   "
> "   Guayaquil or Carthagenia.........6 to 8   "

The last two being harder require less, while the East Indian and African being softer require more sulphur than the Para. These are to be mixed in the usual manner. For developing the best qualities of the compound, he depends upon the proper mode of vulcanizing.

He remarks—"In the vulcanizing process, there is eliminated during the whole operation a constant discharge of sulphuretted hydrogen, and other sulphuretted gases, which must have means of escape through the pores of the mass whilst being vulcanized.

"The escape of these gases from soft elastic rubber is very easy, but from hard rubber or gutta percha, whose pores on the surface portions are very close, it is difficult for them to escape. The consequence is, that the mass is liable to be exploded from the increased pressure of the confined gases within it. Hence, the triple length of time required to vulcanize my composition, and the greater heat to expel the gases. From the greater degree of compactness of this composition, with caoutchouc and sulphur alone in it, great difficulty has been experienced in vulcanizing pieces of half an inch in thickness. But by my present improved management of the heat in vulcanizing (raising it very gradually step by step up to the highest point) I am enabled to vulcanize pieces of an inch or more in thickness with great uniformity and perfection.

"Different rubber compounds containing dissimilar ingredients will not vulcanize in the same time and at the same temperature, but the time and temperature must be adapted to the constitution of the mass or mixture to develop its best qualities. A mixture containing much earthy matter may be vulcanized in much shorter time than one constituted of caoutchouc and sulphur alone, and yet be solid, owing to the earthy matter facilitating the escape of the gases evolved in its substance, at the same time such compositions are destitute of elasticity and flexibility. Suppose the vulcanizing heat be set from 274° to 300°, the earthy base composition would be worthless and brittle at the end of six hours, and nearly charred at the end of eighteen hours heat; while my composition of caoutchouc and sulphur only would have an ivory hardness, with the spring temper of steel.

"Vulcanization is more difficult in thick than with thin articles, from the fact already stated respecting sulphuretted gases escaping through the pores of the gum; under these circumstances if the mass harden externally faster than internally, the confined gases may explode the mass and spoil the form. Therefore the heat is continued a long time at 275°. For a piece that is about five-eighths of an inch thick, the time required for vulcanizing is thirteen and a half hours.

It is first retained at..............................275° 6 hours.
Then raised to and held at.........................280 3 "
    "   "   "   ....290 2 "
    "   "   "   ....295 2 "
    "   "   "   ....300 ½ "

"Experiments with same grades of time commencing at the highest, then lowering the heat, also raising it to 295°, and retaining the whole period, produced unsatisfactory results."

The preceding extracts from the few of the vast number of rubber patents, give but little information as to the composition of the mixtures vended for dental purposes, yet by exhibiting the modes of proceeding in the hardening of the several compounds by experts, they will serve to throw some light upon the subject so as to assist us in producing the best results from the compositions which we use.

To obtain a knowledge of the value of these compounds and also of their composition, numerous experiments, analytical and synthetical, were performed; some of the most interesting will be related.

FOR FIXED MATTER.

My attention was first called to the subject of rubber being loaded with fixed matter in examining a specimen of dark pink rubber by its great specific gravity, it being 2.188. Maker's name not known.

1st. One hundred grains of this rubber was brought to a white heat to expel all volatile and combustible matter; it did not produce much volume of flame, and left a white ash weighing *sixty grains;* taking into consideration the sulphur and vermillion which were volatilized, it would leave but a small percentage of caoutchouc in the composition.

2d. Two specimens of English light pink rubber obtained from different sources. In the fire they gave out considerable volume of smoke and flame, and each left a residue of *forty-eight per cent.* of white ash.

3*d.* A specimen of English Pink Rubber, made by Ash & Sons', London, marked No. 1. Acted similar to the preceding in the fire. This left a residue of *forty-eight per cent.* of a yellowish white earthy matter. The ash retained the original shape of the rubber.

4*th.* Ash & Sons' Pink Rubber, marked No. 1, X. Acted similar to the two preceding in the fire. It left a residue of *forty-seven per cent.* This was principally, if not wholly white oxide of zinc.

5*th.* Specimen of Ash & Sons' Rubber, marked S. P. Treated one hun dred grains. Under the action of the heat, it first melted down like pure caoutchouc, then emitted a copious volume of smoke and flame. Left a residue of a whitish color, weighing *twenty grains*, this appeared to be principally white oxide of zinc.

6*th.* Specimen of Ash & Sons' Black Rubber, (properly brown). One hundred grains of this was treated as the preceding. Its action in the fire was similar to pure caoutchouc, first melting down and then emitting a dense volume of smoke and bright flame.

Residue.—White and dark ash weighing *four grains.*

7*th.* English Red Rubber. This left *six per cent.* of dark ash or cinder.

8*th.* Dieffenbach's Red Rubber. Left a dark ash or cinder of *sixteen per cent.*

9*th.* Red Rubber. Maker's name unknown. This was treated as above. Melted down, then boiled and foamed, giving off a dense volume of smoke, with a strong smell of asphaltum, leaving *four per cent.* of dark ash. Although this rubber left so small a quantity of fixed matter it is evidently of a very inferior quality, and is largely adulterated with asphaltum or some similar inferior substance. I have been informed that it works badly.

10*th.* American Hard Rubber Company's Red Rubber. This produced a copious flame and dense black smoke, and acted much like pure caoutchouc, while the volatile portions were being given off.

Residue.—*Five per cent.* of dark ash or cinder.

11*th.* American Hard Rubber Company's White Rubber. This did not yield a dense smoke or much flame in comparison to unmixed caoutchouc. Left *fifty-one per cent.* of residue, which was white, compact, and appeared to be a fine quality of white oxide of zinc.

12*th.* American Hard Rubber Company's Brown, for dental purposes. This gave off a dense smoke and copious flame first melting down like pure caoutchouc, leaving *near four per cent.* of dark ash.

13*th.* Specimen of my own make of Brown Rubber, composed of

Para Caoutchouc.. 2
Sulphur... 1

Under the action of the heat this acted precisely similar to the last preceding, and yielded *near three per cent.* of a dark ash.

14th. My own make of Red Rubber.

Caoutchouc.. 48
Sulphur... 24
Vermillion.. 36

In the fire acted similar to pure rubber. One hundred grains left an ash-colored residue of *two per cent.*

15th. Ash & Sons' White. Acted similar to No. 11. Left a residue of *fifty-one per cent.* of white oxide of zinc.

Table Showing the Results of the Preceding Experiments.

	Per cent. of fixed matter.	
1. Specimen of Deep Pink.....................................	60	
2. English Pink..	48	White Clay.
3. Ash & Sons' Pale Pink, No. 1.............................	48	" "
4. " " Deep Pink, No. 1, X........................	47	Ox. zn.
5. " " S. P......................................	20	" "
6. " " Black.....................................	4	Dark ash.
7. " " White.....................................	51	White ox. zn.
8. English Red..	6	Dark ash.
9. Dieffenbach's Red..	16	" "
10. Red unknown, (Asphaltum).................................	4	" "
11. American Hard Rubber Company's Red.......................	5	" "
12. " " " " White	51	White ox. zn.
13. " " " " Brown, near......	4	Dark ash.
14. My own Brown, (C. 2, S. 1), near.........................	3	" "
15. " " Red, (C. 48, S. 24, V. 36)......................	2	" "

These experiments show us that the pink and light rubbers for dental purposes are heavily loaded with such foreign matter as white clay and oxide of zinc, and some to the extent of fifty-one per cent. of their weight. Ash & Sons' S. P. is decidedly the best of his light rubbers, containing only twenty per cent. of fixed matter.

Again, Ash & Sons' Black (brown), the American Hard Rubber Company's brown and my own brown, give results, respectively four, near four and near three per cent. of fixed matter. My own I know was made of pure Para caoutchouc and of sulphur; hence from the residues of the two former so nearly approximating thereto, and also from their similarity of texture and appearance after being vulcanized, we must arrive at the conclusion they are of the same composition, and are therefore good and reliable brown rubbers.

When we examine the results of the experiments upon the English Deep Red, that made by the American Hard Rubber Company, and my own red, we find the fixed matter to be six, five and two per cent. respectively My own red was made of pure Para caoutchouc, vermillion and sulphur

The small disparity of fixed matter found in these rubbers may have arisen from the different state of purity of the caoutchouc used in compounding them.

It is evident that the specimens of English Red and of the American Hard Rubber Company's Red, were not loaded with earthy matter or oxides of zinc or lead, for if they were, the clay would have given us a greater percentage of fixed matter. Oxide of zinc is fixed in the fire at a white heat, and if present would have produced a similar result. Oxide of lead would have shown itself by its reduction, and the greater weight of residue.

The conclusion we would naturally arrive at from the results of these experiments is, that the American Hard Rubber Company's Red and the English Deep Red, are the best Red Rubbers offered for dental purposes.

The American Hard Rubber Company's Red was tested first in January, 1864, the same test was repeated with this make of rubber in the market in January, 1865, with precisely the same results, showing a uniformity in the composition.

To ascertain if there was any free mercury in the American Hard Rubber Company's Red rubber as has been asserted, or any evolved by the decomposition of the sulphuret during vulcanization, a bulb was blown at the end of a glass tube; into this red rubber was inserted, the tube was then bent above the bulb in the form of a retort, and the open end drawn out upwards to a capillary point. The bulb containing the rubber was placed in a bath of paraffin and vulcanized for one hour and a quarter at 320° F.; during which time the opposite end was kept cold, to condense the mercury should any come over. The result of this, and several similar experiments was, no trace of mercury could be detected, free sulphur was sublimed and condensed in a small quantity in the cold parts of the tube.

To ascertain if sulphuretted hydrogen is given off during vulcanization, a bulb was blown at the end of a glass tube, this was filled with red rubber, the tube was then drawn out very small from immediately above the bulb, and curved so that the small part when the bulb was in the paraffin bath could be inserted into a vessel beside it.

The bulb was then placed in a paraffin bath, and the curved end of the tube inserted in a vessel containing a solution of acetate of lead. The heat was raised to 320° F., and retained at that point for one hour and a quarter.

The mean results of several experiments conducted in this manner was, that during the first thirty or forty minutes after the heat had attained to 320°, bubbles of sulphuretted hydrogen came over at short intervals, and at the expiration of this time it was evolved in a continuous stream which continued for a few minutes, causing a copious precipitate of sulphide of lead. After this, until the expiration of the hour and a quarter, the gas was only given off sparingly at intervals. This experiment gives us ocular demonstration that this gas is evolved during vulcanization and in large quantities, and conclusively shows that in thick pieces, especially, the heat should be slowly raised, and the rubber should be under strong pressure to ensure a successful result.

COMPOUNDING RUBBER.

Caoutchouc may be mixed with sulphur, and the coloring matter, by being passed repeatedly between steam-heated rollers; or, the caoutchouc may be first reduced to a pulpy or gelatinous state by some one of its solvents, and the sulphur and coloring matter then mixed with it; in either case the sulphur and coloring should be ground extremely fine, and then the whole ingredients thoroughly incorporated together to insure a satisfactory result.

For experimental purposes the latter method of mixing can be readily practiced by any one. Of the solvents, ether deprived of its alcohol, chloroform and bisulphide of carbon are objectionable on account of their expense, and also the operator being compelled to inhale their vapor during the manipulation. Coal naptha, or benzine, are preferable on this account; they readily reduce the caoutchouc to the proper consistency; but after having been mixed, and the solvent evaporated, the rubber is non-adhesive and does not pack well. Oil of turpentine leaves the rubber somewhat adhesive and in a good condition to pack. Therefore, I have found it a better plan to soften the caoutchouc in oil of turpentine, or, in equal parts of coal naptha, or benzine, and oil of turpentine.

In reducing caoutchouc to a gelatinous condition, it requires a large quantity of the solvent in proportion to the gum. This is remedied by introducing into the solvent from five to fifty per cent. of alcohol; in this case the caoutchouc becomes gelatinous, but does diffuse itself through the solvent, thereby leaving much of it after the softened caoutchouc is removed, for future use.

I generally levigate the coloring matter and sulphur in spirits of turpentine, first reducing the coloring matter very fine, then adding the sulphur, and also reducing it very fine, then add a little of the pulpy caoutchouc, mix thoroughly, and proceed in this manner until the whole is incorporated into a perfectly homogeneous mass. When the coloring matter is ground in linseed oil, the caoutchouc may be softened in naptha, or benzine, and it will pack well, as the oil renders it adhesive; but I am inclined to believe that oil, even in a small quantity, injures the hardness and polish of the rubber.

After the materials are well mixed, the mass should be spread on a glass plate with a spatula, and allowed to remain until the solvent has been evaporated.

The apparatus used in making the following mixtures were a muller and glass plate to grind the colors and sulphur, a spatula, broad-mouth bottles, in which to gelatinize the caoutchouc, and window glass, upon which to spread it when mixed. The caoutchouc was the best Para, and the time and temperature in vulcanizing was the same as that for the American Hard Rubber Company's Red Rubber.

To test the combination of caoutchouc and sulphur alone—

A. $\left\{\begin{array}{ll} \text{Caoutchouc} & 48 \\ \text{Sulphur} & 24 \end{array}\right.$

This gave a dark brown rubber, varying shade in different mixtures; it was strong, compact, and tough, and received a fine polish. This color may be toned down to a dark oak by bleaching in alcohol.

B. This experiment was performed with caoutchouc which had not been smoked; this gum was translucent and nearly colorless, having merely a light straw tint. The proportions were the same as for A.

Result.—Color and properties the same as the above, showing that the natural color of hard rubber composed of simply Caoutchouc and Sulphur is a dark brown.

To test the coloring properties of red oxide of iron.—The following formula gave the best results of the many tried:

C. $\left\{\begin{array}{ll} \text{Caoutchouc} & 48 \\ \text{Sulphur} & 24 \\ \text{Red oxide of Iron, (Rouge)} & 36 \end{array}\right.$

Result.—Texture good; color in different mixtures varied from almost black to black red; the color was more on the red when the rouge was ground in oil than when in spirits of turpentine; after exposure in alcohol to the rays of the sun the red was better developed, but even then it was much darker than the Company's red rubber. The sulphur decomposed the oxide of iron forming a dark sulphide thereby destroying its coloring effect.

Vermillion for Producing a Red.—Numerous experiments were tried to ascertain the quantity of vermillion necessary to overcome the natural brown and produce a red color; the following mixture may be set down as the lowest:

D. $\left\{\begin{array}{ll} \text{Caoutchouc} & 48 \\ \text{Sulphur} & 24 \\ \text{Vermillion} & 36 \end{array}\right.$

Some mixtures made according to this formula were darker and some lighter, owing to the different varieties of vermillion used. The shade was made much lighter by bleaching in alcohol. To bring it to a bright red when vulcanized would require much more vermillion, perhaps equal proportions of caoutchouc and vermillion. This formula produced a good strong compact rubber. If not identical in composition with the Company's Red, it so closely resembles it in texture, strength, appearance, and in every particular, it must very nearly approximate thereto.

To Produce a Yellow.—The coloring effect of chrome yellow was tested; it gave a slate color, the chromate of lead being decomposed, setting free the chromic acid, and forming a sulphide of lead, stone ochre, and Naples yellow and the common orpiment of commerce, were tried with no better results. Pure orpiment or king's yellow gave, when bleached, a lemon yellow, when mixed as follows:

$$
\text{E.} \left\{
\begin{array}{ll}
\text{Caoutchouc} & 48 \\
\text{Sulphur} & 24 \\
\text{King's Yellow} & 36
\end{array}
\right.
$$

Although the color produced by this substance was much more satisfactory than any of the preceding, its use is objectionable, because the texture of the rubber was not good, and the king's yellow being sulphide of arsenic is very poisonous.

The following formula gives a good reliable yellow, viz.:

$$
\text{F.} \left\{
\begin{array}{ll}
\text{Caoutchouc} & 48 \\
\text{Sulphur} & 24 \\
\text{Sulphide of Cadmium} & 36
\end{array}
\right.
$$

This requires bleaching to develop the color fully; it is then much better than that produced by orpiment, is more on the orange, the texture of the rubber is good, and its use is not objectionable.

For a Lighter Yellow—

$$
\text{G.} \left\{
\begin{array}{ll}
\text{Caoutchouc} & 48 \\
\text{Sulphur} & 36 \\
\text{Sulph. Cd} & 36 \\
\text{White ox. Zinc} & 12
\end{array}
\right.
$$

The white oxide of zinc toned down the deep yellow to more of a lemon-color, similar to that produced by the orpiment, at the same time the rubber was of good texture.

Experiments to produce a pink and a flesh-color, so far, have not been successful in producing the desired results, yet some of them are worthy of note.

$$
\text{H.} \left\{
\begin{array}{ll}
\text{Caoutchouc} & 48 \\
\text{Sulphur} & 24 \\
\text{White ox. Zinc} & 30 \\
\text{Vermillion} & 10
\end{array}
\right.
$$

When bleached, gave a dark pink, the color not so good as the English; texture close; not so strong as the brown or red.

Variation of the above.

$$
\text{I.} \left\{
\begin{array}{ll}
\text{Caoutchouc} & 48 \\
\text{Sulphur} & 24 \\
\text{White ox. Zinc} & 36 \\
\text{Vermillion} & 10
\end{array}
\right.
$$

Vulcanized brown, after bleaching, it was a shade lighter than the preceding.

The mixture of

```
      ( Caoutchouc............................ 48
      | Sulphur................................. 24
  K.  { White ox. Zinc....................... 48
      | E. Vermillion......................... 10
      ( Sulphide of Cadmium............... 6
```

after bleaching, produced a buff.

Variation of above.—

```
      ( Caoutchouc............................ 48
      | Sulphur................................. 24
  L.  { White ox. Zinc....................... 96
      | Vermillion............................. 5
      ( Sulphide of Cadmium............... 3
```

This produced a lighter shade—a light buff.

To ascertain the effect of white oxide of zinc upon the natural brown of vulcanized rubber, numerous mixtures were made. The best Lehigh white oxide was used.

```
      ( Caoutchouc............................ 48
  N.  { Sulphur................................. 24
      ( White ox. Zinc....................... 36
```

This produced a drab after bleaching—texture good.

```
      ( Caoutchouc............................ 48
  O.  { Sulphur................................. 24
      ( White oxide of Zinc............... 48
```

When bleached gave a light drab of drab of good texture, and in appearance approximates very near to that of the American Hard Rubber Company's White.

```
      ( Caoutchouc............................ 48
  P.  { Sulphur................................. 24
      ( White oxide of Zinc............... 96
```

This after bleaching gave a grayish white. These three preceding mixtures were repeated by varying the proportion of sulphur, substituting thirty-six for twenty-four, the object of this was to give the rubber additional hardness; this change of proportions had the desired effect, but at the same time the color was impaired. All of these mixtures vulcanize a brownish color, and require to be bleached by the rays of the sun in alcohol for their development.

To produce a black rubber.

```
      ( Caoutchouc............................ 48
  T.  { Sulphur................................. 24
      ( Ivory black, or drop black... 24
```

This mixture gave a good black.

```
      ( Caoutchouc............................ 48
  U.  { Sulphur................................. 24
      ( Ivory, or drop black............ 48
```

This produced an excellent *jet black*, the rubber was hard and of good texture.

The drop black which is in lumps containing gum I have uniformly found to produce a porous rubber, whilst the article under the same name found in commerce, free from gum, gave good results.

By taking several of these different mixtures, (such as the taste of the operator may dictate) and cutting them into shreds, then incorporating them together, and again cutting the mass into small pieces suitable for packing, a very pretty mottled rubber may be made, suitable for hurdles, &c.

After being vulcanized and polished, it must be bleached in alcohol to fully develop the colors, although some of the mixtures present a pleasing appearance without the bleaching process.

In finishing mottled rubber, owing to the several colored mixtures having a different degree of hardness, after the file, prepare for the polishing process by obliterating the file marks with a flat piece of Scotch stone.

The introduction of shellac was tried in one experiment, viz.:

$$
\begin{cases}
\text{Caoutchouc} & 48 \\
\text{Sulphur} & 24 \\
\text{Vermillion} & 40 \\
\text{Shellac} & 12
\end{cases}
$$

The addition of shellac did appear to improve the compound in appearance or texture.

I have now presented the most interesting of the successful results of my experiments in compounding mixtures for making hard rubber, and would now call the attention of those who desire to pursue this subject experimentally, that to color rubber, three points are essential: First, the color must remain unchanged at the heat required for vulcanization. Second, it must withstand the action of sulphur at this temperature; and third, sufficient quantity must be added to the mixture to overpower the natural brown of vulcanized rubber, before its shade can be developed. This fact shows us that all highly colored rubbers, or where the brown is widely departed from, must be weakened by their being loaded with so much color or foreign matter; in proof of this, I have found no other mixture possessing strength and toughness equal to that made of simply caoutchouc and sulphur.

The following table gives, very nearly, the percentage of caoutchouc contained in several of the preceding formula. Also that of Ash & Sons' Pink No. 1, their S. P., their white, and the white made by the American Hard Rubber Company. The percentage given of these latter is based upon calculation.

From the results of the preceding experiments, it is evident we may substitute Ash & Sons' black and the American Hard Rubber Company's brown for the A, brown in the table. Also the English deep Red and the American Hard Rubber Company's Red for D, the red in the table.

	Caoutchouc.	Sulphur.	Vermillion			Parts in.
A. Brown	66⅔	33⅓				100
D. Red	44	22	33			99
F. Yellow	44	22			33 (Sulp. Cad.)	99
H. Pink	42⅔	21½	9		27 (Ox. Zn.)	100
K. Buff	35.4	17.2+	7.3	4.4	35.4 (Sulp. Cad.)	100
N. Drab	44	22			33	99
O. Lighter Drab	40	20			40	
P. Grayish White	28.5	14.3			57.1+	100
T. Black	50	25			25 (Black.)	100
U. Jet	40	20			40	100
Ash & Sons' Pink, No. 1	24	12	18		48 (White earthy matter.)	102
" " S. P.	35.6	17.8	26.6		20 (Ox. Zn.)	100
" " White	32⅔	16¼			51	100
Am. H. R. Co.'s White	32⅔	16¼			51	100

The calculation for the component parts of Ash & Sons' Pink Rubber is based upon the method given in the patent for making pink rubber for dental uses, the quantity of fixed matter it is found to contain, and taking formula D as the composition of red rubber. It will be found, upon examination of this data, that if there is any error in the quantity of caoutchouc given to the pink, it is in its favor. A glance at the table will at once show its and other light rubbers' inferiority to either brown or red for dental purposes.

The calculation of the percentage of Ash & Sons' S. P. is based upon the quantity of fixed matter found in it, and that fixed matter having been mixed a red rubber compounded as in formula D. This is evidently superior to the pink, but inferior to either the red or brown.

Caoutchouc being the *cement* which binds the whole together, if any compound should contain but a small proportion of it, and if any substance prejudicial to the system should enter into its composition, (and in the patent referred to for making pink rubber, such substances are recommended), its weakness of texture from the want of sufficient adhesion of its particles would render it liable to produce injurious effects by its susceptibility to abrasion in the mouth.

ARTICLES

VULCANIZING PROCESS.

Whitney's Vulcanizer, for two Flasks, complete......................$15 75

Whitney's Vulcanizer, for three Flasks, complete.................... 16 80

Whitney's Flasks, each... 87

Kerosene Stove, adapted to Whitney's Vulcanizer................... 2 75

Thermometer Tube and Scale... 1 00

Hayes' High Pressure Vulcanizing Oven, one Flask................. 13 65

Hayes' High Pressure Vulcanizing Oven, two Flasks............... 14 70

Hayes' Vulcanizing Boiler, for two Flasks 15 75

Hayes' Vulcanizing Boiler, for three Flasks 16 80

Hayes' Flasks, each 50

Clamps for holding one, two, or three Flasks, each...... 50

Kerosene Burner, adapted to Hayes' Vulcanizer 2 50

Thermometer Tube and Scale... 1 00

American Vulcanizer, with Gas and Spirit Burners and Safety
    Valve, for three Flasks.. 22 00

Star Flasks (page 13, fig. 5), adapted to either Whitney's or the
    American Vulcanizer, Malleable Iron......................... 1 50

Star Flasks, Brass.. .. 2 00

Bolt and Nut for same... 12

Lamps, Gas Burners, Wrenches, Bolts, Packing, etc., for all
    makes of Vulcanizers, sold separately.

Howell's Vulcanite Packer 15 00

NOTE.—For cuts of Vulcanizers, see page 25.

Brush Wheels, Hard and Soft, Straight and Cup-shaped, of various diameters and widths, 170 varieties, each......$0 22 to 92

Felt Wheels, $1\frac{3}{4}$ and $2\frac{1}{2}$ inches diameter, each ....20 and 40

Cork Wheels, Round and Cone Shape, each............................ 5

Corundum Wheels, from $\frac{1}{2}$ inch to 6 inches diameter............07 to 3 00

Burs for Lathe, Vulcanite...75 to 1 38

Pumice, ground, per pound... 15

Prepared Chalk, per pound... 15

Rotten Stone, per box.. 10

Powdered Calcined Buck-Horn, per box...........................12 and 38

Collodion, 2 ounce Bottle and Brush 50

Liquid Silex, per bottle.. 17

Sandarac Varnish, per bottle 25

Callipers, Steel, page 33, fig. 17, each................................. 1 00

Callipers, Brass, page 33, fig. 16, each 50

Scrapers, page 32, fig. 14, each.. 25

Scrapers, page 32, fig. 15, each.. 50

Wax Knives, page 11, fig. 4, each 20

Wax Knives, page 11, fig. 3, each.. 75

Wax Knives, File-Cut Steel Handles, each............................. 40

Packers, 6 in a set, each 25

Rubber Gauge, page 22, fig. 12 .... 1 25

Water Box, holds four Flasks, made of extra heavy tin, page 21, fig. 11 .... 1 00

American Hard Rubber Company's Gum, per pound................ 4 00

American Hard Rubber Company's Gutta Percha, per pound..... 3 00

English Rubber, Pink, No. 1x, per pound............................. 11 00

English Rubber, Pink, No. 1, per pound...... .... 11 00

English Rubber, Pink, S. P., per pound............................... 9 00

English Rubber, White, per pound...................................... 9 00

English Rubber, Black, per pound 4 00

Base Plate Wax, half-pound box... 58

Base Plate Wax and Gutta Percha, half-pound box................. 58

SAMUEL S. WHITE,

528 Arch St., Philadelphia.
658 Broadway, New York.
100 & 102 Randolph St., Chicago.
16 Tremont Row, Boston.

www.ingramcontent.com/pod-product-compliance
Lightning Source LLC
Chambersburg PA
CBHW022010190326
41519CB00010B/1463